Ednaldo Barbosa Pereira Junior
Fábio H. T. Oliveira

Nitrogen and phosphate fertilization in cowpea cultivation

Ednaldo Barbosa Pereira Junior
Fábio H. T. Oliveira

Nitrogen and phosphate fertilization in cowpea cultivation

Plant nutrition with economic and environmental efficiency

ScienciaScripts

Imprint

Any brand names and product names mentioned in this book are subject to trademark, brand or patent protection and are trademarks or registered trademarks of their respective holders. The use of brand names, product names, common names, trade names, product descriptions etc. even without a particular marking in this work is in no way to be construed to mean that such names may be regarded as unrestricted in respect of trademark and brand protection legislation and could thus be used by anyone.

Cover image: www.ingimage.com

This book is a translation from the original published under ISBN 978-613-9-61913-9.

Publisher:
Sciencia Scripts
is a trademark of
Dodo Books Indian Ocean Ltd. and OmniScriptum S.R.L publishing group

120 High Road, East Finchley, London, N2 9ED, United Kingdom
Str. Armeneasca 28/1, office 1, Chisinau MD-2012, Republic of Moldova, Europe
Printed at: see last page
ISBN: 978-620-7-70726-3

SUMMARY

SUMMARY

EDNALDO BARBOSA PEREIRA JUNIOR & FABIO HENRIQUE TAVARES DE OLIVEIRA
Nitrogen and Phosphate Fertilization in the Caupi Bean Crop, Universidade Federal Rural do Semiàrido, Mossoró - RN, 2012. Thesis (Doctorate in Plant Science).

Nitrogen (N) and phosphorus (P) are generally the two nutrients that occur at the lowest levels in the soil in relation to the plant's needs and are two of the nutrients most required by the cowpea crop. The aim of this study was to estimate the best combination of recommended doses of N and P O_{25} , as well as the critical levels of N and P in the plant for maximum economic production of irrigated cowpeas. To this end, an experiment was carried out under field conditions at the Federal Institute of Paraiba, in the municipality of Sousa-PB. The treatments resulted from a combination of four doses of N (25, 50, 75 and 100 kg ha^{-1}) and four doses of P2O5 (25, 50, 75 and 100 kg ha^{-1}), plus a control treatment (zero dose of N and zero dose of P2O5), totaling 17 treatments. The experimental design was randomized blocks with four replications. The cultivar used was "rib-of-cow", planted at a spacing of 0.80 x 0.20 m, under micro-sprinkler irrigation conditions. The N and P contents in the bean's diagnostic leaf, the available P content in the soil after harvest, pod length, dry matter of the vegetative aerial part, mass of 100 grains, grain yield and crude protein content in the grains were evaluated. The statistical analyses consisted of analysis of variance and multiple linear regression (response surface), considering the doses of N and P O_{25} as independent variables. These equations were used to estimate the critical level of available P in the soil and the critical levels of N and P in the plant. The doses of N and P O_{25} associated with the maximum production of the bean plant were estimated and, based on the gross revenue and the price of fertilizers, the doses of N and P O_{25} recommended for the maximum economic production of the irrigated bean plant were estimated. Increasing the doses of N and P O_{25} applied to the soil increased the dry matter production of the aerial part, the mass of 100 grains, grain yield and the levels of N and P in the diagnostic leaf of the bean plant. The critical levels of N and P in the plant were 46.8 and 3.24 g kg^{-1} , respectively. In the soil, the critical level of P extracted using the Mehlich-1 extractor was 48.2 mg dm^{-3} . Applying just 25 kg ha^{-1} of N + 75 kg ha^{-1} of P O_{25} would yield 1,792 kg ha^{-1} of grain, with a net income of R\$ 5,809.11, corresponding to 94% of the estimated maximum net income.

Keywords: bean plant, nitrogen, phosphorus.

1. INTRODUCTION

Cowpea (*Vigna unguiculata* (L.) Walp.), also known as mung bean or string bean, is a species widely distributed throughout the world, mainly in tropical regions, due to their similar soil and climatic conditions to those of its probable cradle of origin: Africa (BRITO et al. 2009).

According to a survey by CONAB (2012), Brazil's production of beans in the 2011/2012 harvest was 2,906.5 million tons, which kept the country as the world's largest producer of the grain. As far as the Northeast is concerned, CONAB's August 2012 harvest monitoring shows that although the region has the largest planted area in the country, with 1,506,900 ha, it only has the fourth largest production, around 296,500 tons, a fact explained by low productivity when compared to other regions: 1,749 kg ha^{-1} was the average productivity in the Midwest region, compared to 196 kg ha^{-1} in the Northeast.

In the state of Paraiba, cowpeas are grown in almost all the micro-regions, mainly by small farmers in the Agreste and Sertào. They have a considerable production, with indices ranging from 300 to 700 kg ha^{-1} , where 75% of the bean-growing areas are cultivated. Thus, it plays an effective role in people's diets, as it is an excellent source of low-cost protein and carbohydrates.

According to the exploratory soil survey of the state of Paraiba, the predominant soils in the Sousa region are Vertissolos (50%), Planossolo Nàtrico (20%), Neossolo Litólico (15%), Fluvial Neossolo (15%) and Argissolo (10%) (Brasil, 1972), with an emphasis on the production of green coconut, banana, guava and mango and annual crops such as caupi beans, corn and rice, grown in the second half of the year under irrigation conditions.

Among the main factors responsible for the low productivity of cowpeas in the region are irregular rainfall and low levels of nitrogen (N) and phosphorus (P) in the soil. Under irrigated conditions, the lack of an adequate nitrogen and phosphorus fertilization recommendation for cowpeas, based on field experiments conducted in the region, has been pointed out by farmers in Paraíba as one of the main limitations to obtaining high yields.

Some work has been done on nitrogen and phosphate fertilization in bean crops (FAGERIA and BALIGAR, 1996; CARVALHO et al., 2003), but little has been done on this subject with cowpeas.

Dutra et al. (2012), evaluating nitrogen fertilization in cowpea cv. Canapuzinho at levels of 15 and 30 kg ha^{-1} , in the foundation or top dressing, did not interfere with the productivity of the plants or the physiological quality of the seeds.

A study carried out by Silva et al. (2010) on yellow latosol found that regardless of how the fertilizer was applied, the recommended dose of P (60 kg ha^{-1} P2O5) for the cowpea crop was not enough to provide the highest yields.

In the state of Paraiba, no studies were found on the response curves of caupi beans to doses of N and $P O_{25}$ in field conditions and under irrigation.

In view of the above, there is a need to carry out research in the region on nitrogen and phosphate fertilization for irrigated cowpeas, with a view to drawing up recommendations for fertilizing this crop with these nutrients. Defining doses of N and $P2O5$ that do not fall short of or exceed the crop's real needs will allow the producer to earn a higher net income and avoid wasting fertilizer that has been over-applied to the soil.

The aim of this work was to estimate the best combination of recommended doses of N and $P O_{25}$, as well as the critical levels of N and P in the plant for the maximum economic production of irrigated cowpeas in semi-arid climate conditions.

2. LITERATURE review

2.1. Bean cultivation in the Northeast

Originally from Africa, the cowpea is believed to have been introduced to Latin America in the 16th century by Spanish and Portuguese colonizers. It arrived in Brazil, probably in the state of Bahia (FREIRE FILHO, 1988), and was taken by the colonizers to other areas of the Northeast and to other regions of the country. The use of the caupi bean is very similar to the common bean, but it adapts better to the climatic conditions of the semi-arid, humid and sub-humid tropics, so it should be considered a complementary crop and not a competitor to the common bean (SMARTT, 1990).

The first area to be cultivated with cowpeas was in Bahia, spreading to all regions. However, its cultivation is concentrated in the North and Northeast regions and it is one of the main social and economic alternatives for food supplements, due to its high protein content and job creation (FREIRE FILHO et al. 2005).

The cowpea (*Vigna unguiculata (L) Walp*) is one of the most important crops in the North and Northeast regions of Brazil, as it plays a fundamental role in the socio-economic context of low-income families living in these regions. It provides food with a high nutritional value, as it has a high protein content and also helps to generate employment and income. Its seeds are a source of protein, amino acids, thiamine, niacin and dietary fiber; therefore, it is an option for public policy programs focused on improving quality of life, especially in growing areas in rural and urban areas (SOUZA, 2005).

The caupi bean, also known as macaçar bean, string bean or fradinho bean, is an important crop in the economy of the Northeast of Brazil and is of great social significance, being the main protein and energy food for rural people. Due to their nutritional value, cowpeas are grown mainly to produce dried or green beans ("feijãoo-verde" with a moisture content of between 60 and 70%) for human consumption, in canned or dehydrated form. It is also used as green fodder, hay, silage, flour for animal feed and as green manure and soil protection (ANDRADE JUNIOR, 2000).

The Northeast of Brazil has the lowest bean yields, estimated for the 2010-2011 harvest at 482 kg ha^{-1} (CONAB, 2011). Productivity in the Northeast and not even in Brazil reflects the productive potential of beans. According to Rosolem and Marubayashi (1994), production systems, climatic effects, crop health and even farmers' economic problems are factors that lead to low yields.

Due to its hardiness, this crop does very well in the north and northeast, where temperatures are high and water availability is low. According to Freire Filho et al. (2005), the species responds well to different levels of stress throughout the stages of its development and is most affected during the grain filling phase.

2.2. Socio-economic importance of cowpeas

Brazil is the world's largest producer of beans, consuming all of its production, but it still imports quantities that complement its demand, making it a net importer of this product. Cowpea is the crop that produces the most in the Northeast, with an area corresponding to approximately 60% of the total bean area. The area harvested, production and productivity fluctuate greatly from year to year, mainly due to climatic variations. Estimates point to production of 3,732,000 tons of beans for the 2010-2011 harvest, with yields of 935 kg ha^{-1} , and 2,906,500 tons of beans for the 2011-2012 harvest, with yields of 889 kg ha^{-1} (CONAB, 2012).

Considering that its average consumption is 20 kg per person per year, it supplies the tables of 27.5 million northeasterners and generates 2.4 million jobs. These figures are extremely important, because they reflect the crop's participation in the context of generating employment, income and food production in the country and make it worthy of greater attention from supply policies and research support bodies (FREIRE FILHO et al. 2005).

The caupi bean is an excellent source of protein (23-25% on average) and has all the essential amino acids, carbohydrates (62% on average), vitamins and minerals, as well as being high in dietary fiber, low in fat (2% oil content on average) and contains no cholesterol. It is a staple food for low-income populations in the northeast of Brazil. It has a short cycle, low water requirements and is hardy enough to grow in low fertility soils (ANDRADE JÙNIOR, 2000).

Cowpeas used to be a traditional crop with a restricted market. In recent years, fortunately, it has gained greater economic expression. It is grown by small, medium and large producers who use high technology, and its market is expanding beyond the borders of the North and Northeast regions. It's worth mentioning that some classes of grain are already being traded on commodities exchanges in the Southeast, such as black-eyed beans (FREIRE FILHO et al. 2005).

2.3. Bean nutrition and fertilization

Beans are considered a nutrient-demanding plant due to their small, shallow root system and short cycle. It is therefore essential that the nutrient is made available to the plant at the right time and place. Although there are disparities in the literature with regard to the quantities of nutrients absorbed by the bean plant, the requirement is usually greater than that of soybeans, for example. The average amounts of nutrients exported per 1,000 kg of grain cited in various studies are: 35.5 kg of N, 4.0 kg of P, 15.3 kg of K, 3.1 kg of Ca, 2.6 kg of Mg and 5.4 kg of S (ROSOLEM and MARUBAYASHI, 1994).

Fageria and Baligar (1996) reported that to produce one ton of grain, bean plants need to extract 23 kg of N, 3.5 kg of P, 22 kg of K, 6 kg of Ca, 3 kg of Mg, 52 g of Zn, 11g of Cu, 121g of Mn, 220g of Fe and 16g of B. Nutrient intakes in the bean crop were in the following order: N>K>Ca>P>Mg, among the macronutrients, and Fe>Mn>Zn>B>Cu, among the micronutrients. Fageria & Santos (1998) also reported the same order of nutrient accumulation in the bean crop, in a varzea soil.

Mineral fertilization plays an important role in the growth and development of crops. Nitrogen and potassium, supplied in a balanced way, promote vegetative growth, the formation of flower and fruit buds (MARSCHNER, 1995), increase resistance to pests and diseases (MALAVOLTA et al. 1989; MARSCHNER, 1995), while phosphorus is essential for photosynthesis, cell division and the development of the root system, as well as promoting abundant flowering and fruiting, directly influencing the productivity and quality of the products harvested (FILGUEIRA, 2000).

Macronutrients are involved in various metabolic processes in plants. Nitrogen is an important and limiting element in crop production, especially in those that provide green mass; phosphorus is indispensable to plant life because it plays a part in cell division, sexual reproduction, photosynthesis, respiration and the synthesis of organic substances; and potassium acts in the protection mechanism and in stomatal control (OLIVEIRA et al. 1996).

In bean plants, the most commonly used critical ranges for leaf contents are those proposed by Wilcox and Fageria (1976), with variations in g kg^{-1} : N - 28.0 to 60.0; P - 2.5 to 5.0; K - 18.0 to 50.0; Ca - 8.0 to 30.0; Mg and S - 2.5 to 7.0 and considering the collection of the first mature leaf from the tip of the branch (30 leaves per hectare), collected at the beginning of the flowering stage of the bean plant, those of Malavolta et al. (1997) with values in g kg^{-1} : N - 30.0 to 50.0; P - 2.0 to 3.0; K - 20.0 to 25.0; Ca - 15.0 to 20.0; Mg - 4.0 to 7.0; S - 5.0 to 10.0, used as critical ranges for assessing the nutritional status of the crop by foliar diagnosis. Also, in Minas Gerais, Martinez et al. (1999) recommend using the following reference values for interpreting the results of the analysis of leaves sampled from the middle third of the bean plant in g kg^{-1} of dry mass: for N - 30.0 to 35.0; P - 4.0 to 7.0; K - 27.0 to 35.0; Ca - 25 to 35; Mg - 3.0 to 6.0; S - 1.5 to 2.0.

2.4. Nitrogen fertilization

Nitrogen is a primary macronutrient, essential for plants because it plays a part in the formation of proteins, amino acids and other important compounds in plant metabolism. Its absence blocks the synthesis of cytokinin, the hormone responsible for plant growth, causing a reduction in size and consequently a reduction in the economic production of seeds (MENGEL & KIRKBY, 1987).

Carvalho et al. (1999) found that nitrogen sources and application methods influenced the physiological quality of the seeds. However, Paulino et al. (1999) found no significant differences between nitrogen sources and ways of spreading nitrogen on the physiological quality of bean seeds.

Bassan et al. (2001), studying the cultivar Pérola "in winter", found increasing germination values of over 90% with the application of nitrogen up to the dose of 90 kg ha^{-1} of N in top dressing, in the absence of foliar fertilization with molybdenum. These authors also reported that the dose of 58 kg ha-1 of N allowed the maximum germination value for the accelerated ageing test (81%) in the treatment referring to the application of cover nitrogen fertilization without inoculation.

Carvalho et al. (2001) found no positive effect of nitrogen doses and application times on germination and vigor (accelerated aging) for the IAC Carioca cultivar, "in winter". Crusciol et al. (2003), in a study carried out during the "spring" period with this cultivar, also found no significant effect of nitrogen doses, either when sowing or when top dressing, on germination, which was above 90%.

The amount of Nitrogen (N) supplied by most soils is small. Very little is found in rocks and minerals; much of the N in the soil comes from organic matter. Organic matter releases N slowly, and the rate is controlled by factors such as temperature, humidity and texture. In general, around 20 to 30 kg of N per hectare is released annually for every 1% of organic matter contained in the soil. Thus, a soil with 2% organic matter could release 40 to 60 kg of N per year. One of the products of organic matter decomposition is ammonium, which can be retained by the soil, absorbed by plants or converted into nitrate. Nitrate can be used by plants, leached out of the root zone or converted to gaseous N and lost to the atmosphere (ALVES, 2006).

Applying N as a top dressing to irrigated bean plants, Barbosa Filho et al. (2001) concluded that there is no difference between applying urea fertilizer and ammonium sulphate on the surface or incorporated into the soil. Applying urea fertilizer to the surface of the soil followed by irrigation is the most economical option for top dressing irrigated beans. It is recommended to apply 120 to 150 kg of N, with half applied at 15 and the rest at 30 days after emergence, by surface application to the soil followed by irrigation or via irrigation water, using urea fertilizer as the N source.

Nitrogen fertilizer management differs from the management of other nutrients because decision-making involves technical, economic and environmental aspects (Ceretta & Silveira, 2002), since this nutrient is subject to losses through erosion, leaching, denitrification and volatilization (Amado et al. 2002).

In rainfed crops, the probability of a response in beans is lower, as is the production potential. However, in irrigated crops, in addition to greater production potential, there is better use of the fertilizer applied, making nitrogen fertilization essential. It is usually recommended to apply 1/3 of the N dose at sowing, and 2/3 should be applied up to 20 days after crop emergence. With high doses of N, top dressing could be applied in up to two installments, the first between 15 and 20 days and the second up to 35 days after plant emergence. There are no research results confirming this recommendation, but the study of N absorption patterns and an understanding of the nutrient's functions and effects on the plant allow this inference to be made (ROSOLEM & MARUBAYASHI 1994).

Protein content is directly linked to nitrogen, an element of great importance and made available to plants by the addition of mineral nitrogen fertilizer to the soil, by the addition of organic matter or by the fixation of nitrogen from the air by microorganisms (RAIJ, 1991).

Nitrogen is one of the nutrients that provides the greatest response to common beans with doses above 100 kg ha^{-1} (VIEIRA, 1983). Maia et al. (2005) state that the protein content of beans can be altered by the fertilizer used, mainly by the nitrogen content of the fertilizers. Especially in those that provide green mass (OLIVEIRA et al. 1996).

Andrade et al. (2004) found that the protein content in bean grains was directly related to nitrogen fertilization, using three different doses of the fertilizer, and the beans produced with the highest dose had the highest protein content, while the lowest dose was responsible for producing grains with the least protein.

Soratto et al. (2004) observed that in the conventional tillage system, in succession to the maize crop, the maximum productivity of the bean plant was achieved with an estimated dose of 129 kg ha^{-1} of N in top dressing, while in the no-till system the estimated dose for maximum productivity was 182 kg ha^{-1} of N, indicating a greater demand for the nutrient in this system.

Nitrogen fertilization in the cultivation of common beans in tropical lowlands provides the producer with considerable economic gains due to the plant's response to the nutrient, where the application of part of the nitrogen incorporated into the soil was more effective than the application of the nitrogen applied to the surface (SANTOS et al. 2009).

Stone and Moreira (2001) found that the number of pods per plant, the mass of 100 seeds and the productivity of bean plants responded significantly to the use of N, applied 35 days after emergence, under the no-till system. They also found that there was an increase in productivity over several years of cultivation as the doses of this nutrient increased.

Beans are a nutrient-demanding plant, among which N is absorbed in the highest quantities. According to Oliveira et al. (1996), a quantity of more than 100 kg ha^{-1} of N is required to guarantee the extraction of the nutrient associated with high yields.

Recent studies show that the supply of N through mineral fertilization affects the process of biological nitrogen fixation in legumes, Oliveira et al. (2003) says that plants can directly absorb the N present in the soil, as it is in a more accessible form. Studies carried out by Xavier (2006) found that increasing doses of nitrogen reduced nodulation in cowpea plants and there was no significant increase in dry matter accumulation.

Almeida et al. (1988) found that an increase in N doses led to an increase in the ratio between the aerial part and the root system, thus affecting the movement of carbohydrates within the plant and consequently leading to an imbalance between photosynthesis and respiration.

The most widely used sources of N in Brazilian agriculture are urea and ammonium sulphate. Due to its characteristics and reaction in the soil, urea has a high potential for losing NH_3 through volatilization (KELLER & MENGEL, 1986; LARA & TRIVELIN, 1990) and ammonium sulphate, in addition to the possibility of losing NH3, has a high capacity for acidifying the soil (Barbosa Filho et al. 2001).

In his research, Smiderle (2004) found that as the dose of N increased, there was a reduction in the germination of cowpea seed, although the seed produced had a higher nutritional quality. This conclusion contradicts the results found by Soratto et al. (1999), who obtained a linear increase in seed germination and an improvement in vigor when applying nitrogen top dressing to common beans.

The effect of applying nitrogen top dressing doses to common beans (0, 25, 50, 75 and 100 kg ha^{-1}), under different tillage systems (conventional, minimum and direct) was tested by Silva et al. (2004) on full bloom, plant dry matter, number of pods and grains per plant, number of grains per pod, mass of 100 grains, cycle, grain yield and plant nitrogen content, where they found that grain yield was influenced by nitrogen doses and significant increases were obtained with the application of 75 to 100 kg N ha^{-1} .

Oliveira et al. (2003), in the Areia/PB region, found maximum estimated yields of pods (11 and 10 t ha^{-1}), green grains (9.3 and 8.4 t ha^{-1}) and dry grains (3.55 and 3.44 t ha^{-1}) obtained by using nitrogen applied to the soil and via foliar application, respectively. The author also reports that nitrogen supplied to the soil was more efficient for the cowpea to express its maximum yield capacity.

In a conventional tillage system, Silva et al. (2000) obtained a quadratic response of the bean plant to N, and maximum productivity was achieved with 74 kg ha^{-1} of this nutrient. In the same type of soil, under a no-till system, Soratto et al. (2001) found linear yield responses up to the maximum dose tested, i.e. 100 and 150 kg ha^{-1} , respectively.

Crusciol et al., (2007) applied doses (0, 30, 60 and 120 kg ha^{-1}) of N in cover crops, in the form of nitrocalcium, which promoted greater absorption of nitrate, K, Ca and Mg by the bean plant grown in the no-till system, compared to the application of urea. Top dressing nitrogen fertilization led to an increase in productivity, 100-grain mass and dry matter of the bean crop grown under the no-till system, in succession to black oats, up to the estimated dose of 95 kg ha^{-1} of urea.

Soratto et al. (2001) observed that the application of N as a top dressing at 15, 25 and 35 days after emergence provided better development and increased grain yields under irrigated cultivation in a no-till system.

Meira et al. (2005) reported that nitrogen increased grain yield and this was correlated with the number of pods per plant, with the recommended dose of nitrogen in top dressing being 164 kg ha^{-1}, regardless of the time of application. On the other hand, Rapassi et al. (2003) tested 20, 40, 60, 80 and 100 kg ha^{-1} of N with two sources, urea and ammonium nitrate, in the no-till system, and found that there were no differences between yield levels as a function of the doses of N applied.

Table 1 - Nitrogen and phosphate fertilization recommendations for cowpea cultivation in different Brazilian states.

Recommendation	N			P2O$_5$[0]
	Planting	Coverage	Total	Planting
	--- kg ha^{-1} ----			---- kg ha^{-1} -----
PE [1]	15	20	60	30 - 70
EC[2]	20	-	20	30 - 110
Freire Filho[3]	0	20	20	20 - 60
MG[4]	30	40	70	50 - 90
Average	14	40	54	51

Source:[1] Universidade Federal de Pernambuco (2008); [2] Universidade Federal do Ceara (1993);[3] Freire Filho (2005);[4] CFSEMG (1999).

Nascimento et al. (2004), applying nitrogen topdressing doses (0, 30, 60, 90 Kg ha^{-1}), observed an increase in the nutrient content in the leaves at both stages of the bean crop's development. It can also be inferred that the dose of 90 Kg ha^{-1} of N was the one that provided the highest levels in the leaves in the two seasons evaluated.

2.5. Phosphate fertilization

Phosphorus (P) is a vital component in the process by which plants convert solar energy into food, fiber and oil. P plays a key role in photosynthesis, sugar metabolism, energy storage and transfer, cell division, cell growth and the transfer of genetic information. In addition, this element promotes the initial formation and development of the root, plant growth; accelerates soil cover to protect against erosion; affects the quality of fruit, vegetables and grains, and is vital for seed formation

(MALAVOLTA et al. 1997).

The vast majority of Brazilian soils are acidic, have low fertility and a high phosphorus retention capacity, which leads to the need to apply high doses of phosphates, contributing to an increase in production costs and a reduction in the non-renewable natural resources that produce these inputs (MOURA et al. 2001). In order to achieve high productivity, phosphate fertilization is necessary, which has led to an intensification of the search for the most appropriate doses for the crops and which enable greater economic returns (FAGERIA, 1990).

Phosphorus is the nutrient that has provided the greatest and most frequent crop responses. However, its low availability in the soil negatively affects plant growth and production (PASTORINI et al. 2000). In the common bean crop, Miranda et al. (2000) found that the higher the phosphorus levels, the higher the yields, since there was a linear response to doses of phosphate fertilizer applied at the same time before planting, of up to 1,000 kg ha^{-1} of P2O5 .

Among the various nutrients that plants need, P occupies a prominent place due to its deficiency in the vast majority of our soils. Generally, the P content in the soil varies between 0.2 and 5.0 g kg^{-1} , but its availability to plants is limited due to the form in which it is inorganically fixed and microbially immobilized (ARAÙJO & MACHADO, 2006).

According to Korndorfer et al. (1999), phosphorus sources can basically be divided into soluble, poorly soluble and insoluble. The former, when added to the soil, rapidly increase the concentration of phosphorus in the soil solution. Soluble phosphates decrease in efficiency over time due to the process of "adsorption" or "fixation" of P. Natural phosphates, on the other hand, which are insoluble in water, dissolve slowly in the soil solution and tend to increase the availability of P for plants over time.

Although phosphorus is not required in high quantities like the other macronutrients, it is the limiting nutrient for crop production. The requirement for cowpeas is 60 kg ha^{-1} (Freire Filho 2005). Phosphorus dosage should be based on soil analysis. According to Parry et al. (2008), the availability of other macronutrients is influenced by the presence of phosphorus, since at higher doses of phosphorus there was a higher concentration of other nutrients in the plant.

Phosphorus is the element that most often limits crop production because it is in forms that are not readily available to plants and because of the high adsorption characteristics of soils. Although it is required in small quantities by most crops, large amounts of phosphorus (P) have been applied to meet crop needs (CARVALHO et al., 1995).

Rosolem & Marubayashi (1994) report that bean plants have responded to applied phosphorus in the vast majority of experiments. Under irrigation conditions, higher yields are obtained with the same dose of P, because the fertilizer applied, as well as the P in the soil, is better used by the plant. However, due to the higher yields obtained under these conditions, the economic dose of P will probably be higher than for rainfed crops.

Low P availability is the main limitation to plant production in natural or agricultural ecosystems (Lynch & Brown, 2001). The optimum growth requirement for P is in the range of 0.3 - 0.5 % of the plant's dry phytomass during the period between emergence and flowering (MARSCHNER, 2002).

The P added to the soil by crop fertilization represents a significant variable cost, since much of it is retained in the soil in different forms. The residual effect of the nutrient is likely to be an important contributor to the efficiency and economy of phosphate fertilization. In a planting system with phosphate fertilizers, when previous crops are adequately fertilized, the residual effects of phosphate fertilizers are noticeable. Studies on soils with a high capacity for obtaining P have shown that when they were properly treated with phosphate fertilizers, part of the nutrient remained in the soil and was available to the plants for several crops (AZEVEDO et al. 2004).

Phosphate fertilizer added to the soil, in addition to its immediate effect on the crop that follows the fertilization, can have a residual effect on subsequent crops. In addition to the type of crop, several factors can affect the residual effect of phosphate fertilizers, such as: doses and sources of P, method of application, management, temperature, soil type, time of application and soil moisture. The residual effect of phosphorus has been evaluated by several authors on the production, dry matter yield and P content of subsequent crops (MOREIRA et al. 2002).

Cravo & Smyth (1991) consider P to be the most limiting element for crop development, although it is deficient in 90% of the region's soils, followed by K and N. The way in which the set of processes and reactions affect all the nutrients in the soil-plant system will determine the maintenance of life and the growth of the constituents of its metabolism.

For cowpeas, phosphorus has provided frequent responses and its low availability in the soil negatively affects plant growth and production (PASTORINI et al. 2000). However, although it is the nutrient that crops respond to the most, little is known about the quantities to be used in order to obtain satisfactory yields for cowpeas. The little information on the use of phosphorus in this species reports that in soils with low fertility, it should be applied at planting, together with organic matter (FILGUEIRA, 2000).

For the common bean, of which the cowpea is a differentiated form, phosphorus is the nutrient that has provided the greatest and most frequent responses, and its low availability in the soil negatively affects plant growth and production (PASTORINI et al. 2000).

When P is in the solution, it moves by diffusion to the surface of the roots (HORST et al. 2001). As it is adsorbed, its concentration on the surface of the roots decreases, making it necessary to replenish it. In this sense, root density is very important in the process of P absorption, given that the phosphate anion moves over short distances. As absorption occurs, a concentration gradient of this element is generated in the rhizosphere, which is the driving force for the diffusion of phosphorus to the roots (HINSINGER, 2001).

Fageria and Santos (1998), evaluating phosphate fertilization in a vàrzea soil, found that the P content of 0 to 5.3mg kg^{-1} is classified as very low and, in order to meet this level (5.3mgkg^{-1}), it is necessary to apply 350kgha P O$^{-1}_{25}$ in the furrow. To obtain maximum production at this level, it is advisable to apply 150 kg ha^{-1} P O$_{25}$ in the furrow. To increase the P level from 5.3 to 7.1, you need to apply 560 kg ha P O$^{-1}_{25}$, which is considered a low level to obtain 70 to 90% relative yield. The maximum yield, after meeting this level, can be obtained by applying 100 kg ha P O$^{-1}_{25}$ in the furrow. When the P level is medium (7.1 to 9.0mg kg^{-1}), it is a good idea to apply 760 kg ha P O$^{-1}_{25}$. After reaching this level, it is necessary to apply 100 kg ha P O$^{-1}_{25}$ to obtain maximum bean production. When the P content in the soil reaches 9mg kg^{-1}, it is essential to apply 50kg ha^{-1} P2O5 in the furrow to obtain the maximum yield.

In cowpeas, Oliveira et al. (2005), studying the effect of phosphate fertilization, achieved yields of 30.13 Kg ha^{-1} of pods using a dose of 252 kg ha^{-1} P2O5 and Pôrto et al. (2005) of 17.54 Kg ha^{-1} of pods with a residual dose of 165 kg ha^{-1} of P2O5. Silva et al. (2010) evaluated the effect of five doses 0, 20, 40, 80 and 160 kg of P O$_{25}$ per hectare and two sources of soluble P - simple superphosphate (SFS) and triple superphosphate (SFT) - on cowpeas. Regardless of the sources of P, doses of between 60 and 80 kg of P2O5 ha^{-1}, in sandy soils, provided greater growth for this crop.

Oliveira et al. (2004), evaluating the effect of phosphorus doses on fava bean production, achieved green (5.2 t ha^{-1}) and dry (2.7 t ha^{-1}) grain yields using 309 and 302 kg ha^{-1} of P2O5, respectively.

Silva (2007), applying increasing doses of P2O5 to sandy-textured soil, concluded that with a P content above 141 mg dm^{3} and residual P above 45.20 mg dm^{3}, it should not be recommended to supply this nutrient to cowpeas when dry grains are to be harvested. Leaf tissue P levels of 5.43 g kg^{-1} were achieved as a function of the doses of P O$_{25}$ and with P residue in the soil in the range of 8.53 mg dm^{-3} in cowpeas.

3. MATERIALS AND METHODS

3.1. Characterization and preparation of the experimental area

The thesis work was carried out under field conditions in an area at the Federal Institute of Education, Science and Technology of Paraiba - Campus - Sousa (IFPB), located in the Irrigated Perimeter of São Gonçalo, municipality of Sousa, PB, at an altitude of 264 m, with a southern latitude of 6° 45' and a western longitude of 38° 13', in a soil classified as Planossolo, with a flat relief and a topsoil texture classified as sandy loam.

The climate is characterized as hot semi-arid of the BSH type according to the Koppen classification, meaning that evaporation is greater than precipitation. The average annual rainfall is 654 mm, with rainfall concentrated between January and June. The average temperature is 28 °C, while the relative humidity is 64%. No rainfall was recorded during the period of the experiment.

The experiment was conducted from September to December 2010. To set up the trial, soil samples were taken from a depth of 0-20 cm for chemical and physical analysis in the soil analysis laboratory at the aforementioned teaching unit, as recommended by EMBRAPA (1999). The physical and chemical attributes of the soil in the experimental area, when the experiment was set up, can be seen in Table 2.

Soon after the results of the analysis were confirmed, the soil was prepared with two cross harrows, followed by furrowing. The area corresponding to the trial was demarcated and staked out to facilitate the application of the treatments, planting and irrigation system.

Table 2 - Chemical attributes and clay content of the soil in the experimental area before the experiments, evaluated in the 0-20 cm layer

pH	M.O	P	K^+	In^+	Ca^{2+}	Mg^{2+}	Al^{3+}	(H+Al)	Clay
$H2O$	$g\ kg^{-1}$	$mg\ dm^{-3}$			---- $cmol_c\ dm^{-3}$ -				$g\ kg^{-1}$
7,6	9,26	18	0,66	0,18	4,	12,2	0,0	9,26	152

3.2. Experimental design

Seventeen treatments were applied in a randomized block design with 4 replications, giving a total of 68 plots. The treatments were the result of a factorial combination of four doses of N (25, 50, 75 and 100 kg ha^{-1}) and four doses of P2O5 (25, 50, 75 and 100 kg ha^{-1}), plus an additional treatment (control), as described in Table 3. Each block measured 272 m^2 (54.4 x 5.00 m) and each plot measured 16 m^2 (3.2 x 5.00 m), with the plot consisting of four rows of plants measuring 5 m in length. The beans were planted at a spacing of 0.8 x 0.2 m. The useful area of the plot was formed by the two central rows made up of 46 plants, discarding one plant at each end.

3.3. Application of treatments and crop fertilization

Once the area was ready, planting fertilizer was applied to the bottom of the furrow two days before planting. In this fertilization, fractions of the doses of N, $P O_{25}$, $K_2 O$, B, Zn and Cu corresponding to 20, 100, 50, 100, 100 and 100% of the total doses of these nutrients were applied (Table 3). The sources used for these nutrients were: urea, simple superphosphate, potassium chloride, boric acid, zinc sulphate and copper sulphate.

Two top dressings were applied at 15 and 30 days after plant emergence, one day after weeding to control weeds. In the first top dressing, 40% of the N dose was applied plus the rest (50%) of the K_2O dose (Table 3). In the second top dressing, the remainder (40%) of the N dose corresponding to each treatment was applied (Table 3). In these top dressings, the fertilizers were applied at a depth of about 5 cm and at a distance of 10 cm from the plant and then covered with soil.

Table 3 - Doses of nutrients applied in each treatment of the experiment on the response of cowpeas to doses of N and $P O_{25}$.

Treatment	N	P2O5	K2O	B	Zn	Cu
				- kg ha $^{-1}$ -----		
Witness	0	0	50	1,0	1,0	0,5
1	25	25	50	1,0	1,0	0,5
2	25	50	50	1,0	1,0	0,5
3	25	75	50	1,0	1,0	0,5
4	25	100	50	1,0	1,0	0,5
5	50	25	50	1,0	1,0	0,5
6	50	50	50	1,0	1,0	0,5
7	50	75	50	1,0	1,0	0,5
8	50	100	50	1,0	1,0	0,5
9	75	25	50	1,0	1,0	0,5
10	75	50	50	1,0	1,0	0,5
11	75	75	50	1,0	1,0	0,5
12	75	100	50	1,0	1,0	0,5
13	100	25	50	1,0	1,0	0,5

14	100	50	50	1,0	1,0	0,5
15	100	75	50	1,0	1,0	0,5
16	100	100	50	1,0	1,0	0,5

3.4. Planting the crop and conducting the experiment

Sowing was carried out on September 2, 2010, using the cowpea cultivar *(Vigna unguiculata)* called "rib-of-cow", which has an indeterminate growth habit and a semi-trailing size (TORRES et al. 2008). The spacing used was 0.8 m between rows and 0.2 m between plants, placing two seeds every 0.20 m in the furrow and then covering them with a layer of soil of approximately 5 cm.

The plants emerged five days after sowing. At 10 days after emergence (DAE), the plants were thinned with scissors, leaving just one plant every 0.20 cm of furrow.

During the course of the experiment, two manual weeding sessions were carried out with a hoe to control weeds, always one day before top dressing. To control aphids (*Aphis gossypii*), green leafhoppers (*Empoasca kraemeri*) and whiteflies (*Bemicia tabaci*), the products Dimethoate (dose of 20 ml / 20 liters of water) and Thiamethoxam (dose of 100g/ha) were used.

Two irrigations a day (morning and afternoon) were carried out using a micro-sprinkler irrigation system, in an attempt to provide enough water for the crop to develop properly, so that no problems with water deficit were observed in the field during the course of the experiment.

3.5. Variables analyzed

a) Pod length

The lengths of 10 pods were taken from the three harvests carried out in each treatment, using a ruler graduated in centimeters.

b) Dry matter of the vegetative aerial part

At the end of the experiment, 10 plants were collected, placed in paper bags, duly identified and dried in an oven at 65° for 72 hours until they reached a constant weight.

c) 100-grain pasta

The grains from the three harvests carried out on each plot were mixed and the mass of 100 grains was obtained using a semi-analytical scale.

d) Grain yield

The dry pods were harvested three times (at 68, 75 and 87 DAE) and then dried in the sun. After threshing the pods by hand, the grains were weighed and then the yield was calculated, with the data transformed into kilos per hectare, at 13% humidity (BRASIL 2009).

e) Nitrogen and phosphorus content in the diagnostic leaf

At 46 DAE, when more than half of the plants in each plot were in flower, 20 mature leaves from the tip of the branch were collected from 20 plants in each useful plot (Malavolta et al., 1997). These leaves were dried in an oven with forced air circulation at 65 °C, ground in a Wiley mill and analyzed for N and P content according to the methods described in Tedesco et al. (1995). All these plant analyses were carried out at the UFERSA Soil Fertility and Plant Nutrition Laboratory.

f) Nitrogen and phosphorus levels in the soil

At 47 days after emergence (DAE), composite soil samples were collected at a depth of 0 to 20 cm, using a Dutch auger in the useful area of each plot. To compose the composite sample, two single samples were taken in the planting furrow, four single samples 10 cm from the furrow and six single samples at the midpoint between the furrows, according to the recommendations of Oliveira et al. (2007).

These composite soil samples were analyzed for available P content using the Mehlich extractor (EMBRAPA, 1997) and total N content (TEDESCO et al. 1995).

g) Crude protein content in grain

The crude protein content of the grains was determined by breaking down the protein and other nitrogenous components in the presence of hot concentrated $H_2 SO_4$, according to the Semimicro-Kjeldahl method, multiplying the total N value by a factor of 6.25 (AOAC, 1995). The results were expressed as a percentage. These analyses were carried out at the IFPB-Sousa Physico-Chemical Food Analysis Laboratory.

3.6. Statistical analysis

The statistical analyses consisted of analysis of variance and multiple linear regression (response surface). A multiple linear regression model was fitted to the averages of each treatment, considering the doses of N and P O_{25} as independent variables:

$$Y = a + bN + cN^2 + dP + eP^2 + fNP$$

where Y is the dependent variable, N the doses of Nitrogen (kg ha⁻ 1) and P the doses of P2O5 (kg ha-¹). After fitting this complete model, the coefficients with significance greater than 10% were discarded, and a new, simpler model was fitted with only the parameters that made a significant contribution to the model.

These analyses were carried out using the SAEG software and, when choosing the most appropriate models, the significance of the model parameters was considered using the "t-test", using the mean square of the residual from the analysis of variance of the experiment as the true experimental error (RIBEIRO JÛNIOR, 2001). Response surfaces were then drawn for each characteristic and a significant multiple linear regression model was fitted.

3.7. Economic analysis of fertilization

Using the regression equation with grain yield as the dependent variable of the doses of N and P O_{25} applied, grain yields were estimated for different combinations of doses of Ne and P O_{25} applied. Next, the gross revenue, the costs of nitrogen and phosphate fertilizers and the net revenue were calculated.

When calculating gross revenue, the price of a 60 kg bag of caupi beans in the local trade in Sousa-PB, surveyed in August 2012, was R$200.00 a bag (R$3.33 per kilo of beans). Considering that at that time the price of a 50 kg bag of simple superphosphate was R$ 57.00 and a 50 kg bag of urea was 86.50, it was estimated that 1 kg of P2O5 via simple superphosphate cost R$ 1.14 and 1 kg of N via urea cost R$ 1.73. Based on this information and the values of the doses of N and P applied, the fertilizer costs were calculated. The net income was calculated as the difference between the gross income and the costs of the urea and simple superphosphate fertilizers.

4. Results and discussion

4.1. Pod length, dry matter of the vegetative aerial part and mass of 100 grains

It can be seen (Table 4) that there was no significant effect for pod length and no regression model fitted the observed data. Even though the values of 24.9 cm (25 kg ha^{-1} of N + 100 kg ha^{-1} of P O_{25}) were not significant, there was a 10% increase when compared to the average value of 22.8 cm (control) and 24.5% above the level corresponding to the commercial standard for cowpeas, which is 20 cm (SILVA & OLIVEIRA 1993; MIRANDA et al.1996).

Evaluating new caupi varieties for the Brejo Paraibano micro-region, Santos et al. (2009) observed that the pod lengths of the varieties IPA-207 (22.0 cm), IPA-206 (21.1 cm), EPACE-10 (20.7), BR 17 Gurguéia (19.0 cm), BRS Maratoă (19.0 cm), Costela de vaca (22.3 cm), Cariri (20.0 cm) and canapu (19.8 cm) were lower than in this study.

The data obtained in the present study corroborate those found by Santos et al. (2007) who, with the application of increasing doses of nitrogen, also found no significant difference in the pod length of cowpeas in sandy-textured soil in Lagoa Seca-PB. However, they disagree with those obtained by Silva (2007) who, studying the initial application of P $_{25}$ to the soil in successive crops, had a significant influence ($p<0.01$) on the pod length of cowpeas in a Neossolo Regolitico in the municipality of Areia-PB.

Table 4 shows the results for the dry matter of the vegetative part. There was a significant effect at the 5% probability level for the doses of N and P O_{25} , but no regression model fitted the data obtained. The average values ranged from 1,259 kg ha^{-1} (control) to 2,332 kg ha^{-1} (25 kg ha^{-1} of N + 100 kg ha^{-1} of P O_{25}), showing that the minimum dose of N and the maximum dose of P O_{25} had an efficient influence on this variable. Even so, this maximum dose of 100 kg ha^{-1} of P O_{25} becomes economically viable when

Table 4 - Pod length, dry matter of the vegetative part and mass of 100 grains, as a function of doses of nitrogen and phosphorus applied to the soil

Dose of P O_{25} (kg ha^{-1})	N dose (kg ha)$^{-1}$ 0	25	50	75	100	Average
Pod length (cm) ------------------------------						
0	22,8	-	-	-	-	-
25	-	22,9	23,3	23,4	22,6	**23,0**

50	-	23,6	23,7	23,2	23,1	**23,4**
75	-	23,4	22,6	24,0	23,7	**23,4**
100	-	24,9	23,3	23,8	23,4	**23,8**
Average	-	**23,7**	**23,2**	**23,6**	**23,2**	23,4

ANAVA: CV (%): 4.4 $\qquad$ Ftrat: 1.17^{ns}

Regression: No model fitted the data

Dry Matter of the Vegetative Aerial Part (kg ha) $^{-1}$----------------------------------

0	1.259	-	-	-	-	-
25	-	1.461	1.708	1.788	2018	**1.743**
50	-	1.406	1.246	1.872	1744	**1.567**
75	-	1.708	1.720	1.707	1852	**1.746**
100	-	2.332	1.818	1.994	2009	**2.038**
Average	-	**1.727**	**1.623**	**1.840**	**1.905**	1.773

ANAVA: CV (%): 23.6 $\qquad$ Ftrat: 1.87^{*}

Regression: No model fitted the data

-- Mass of 100 Grains (g) --------------

0	24,2	-	-	-	-	-
25	-	27,4	27,9	29,7	27,9	**28,2**
50	-	29,3	29,3	28,0	28,3	**28,7**
75	-	27,6	27,8	28,7	28,9	**28,2**
100	-	28,1	28,0	28,5	27,3	**27,9**
Average	-	**28,1**	**28,2**	**28,7**	**28,1**	28,2

ANAVA: CV (%): 4.9 Ftrat: 3.03**

Regression: $Y = 24.66 + 0.0726*N - 0.0005*N^2 + 0.06719*P - 0.0005*P$ $R^{22} = 0.70$

$** = P< 0.01; * = P<0.05;^{ns} =$ not significant

compared to what was recommended by Fageria & Santos (1998), who, when evaluating phosphate fertilization on bean plants, found a maximum dry matter production of around 2,736 kg ha^{-1} , with the application of 800 kg ha^{-1} P O_{25} at the same time. After this dose, yields decreased.

The data found in this work do not agree with those obtained by Silva et al. (2002) who, when applying increasing doses of N as a top dressing to bean plants, obtained the highest dry matter values with the application of 100 kg ha^{-1} of this nutrient.

With regard to the mass of 100 grains, the results showed a significant effect at 1% probability for the doses of N and P2O5 applied. For the same variable, there was a range from 24.2 (control) to 29.7 g (75 kg ha^{-1} of N + 25 kg ha^{-1} of P2O5). The regression equation model that best fitted the doses of N and P O_{25} applied was quadratic, with the maximum 100-grain mass estimated at a dose of 66.94 kg ha^{-1} of N and 60.54 kg ha^{-1} of P O_{25} (Table 4, Figure 1).

Work carried out by Soratto et al. (2006) found that doses of N in top dressing had a linear influence on grain mass, increasing as the dose of N applied to the soil increased.

Valderama et al. (2009) found no significant effect for the variable mass of 100 grains, depending on the sources and doses of nitrogen in top dressing (0, 40, 80 and 120 kg ha^{-1}) and phosphorus (0, 50, 100 and 150 kg ha^{-1} of P2O5) in bean plants, with weights varying between 18.25 g and 19.25 g between treatments in the southeast of Brazil.

4.2. Grain yield, nitrogen and phosphorus content in the leaf.

According to the statistical analysis, there were significant effects at the 1% probability level for grain yield between the treatments studied (Table 5). The maximum grain yield was 1,977 kg ha^{-1} (100 kg ha^{-1} of N + 100 kg ha^{-1} of P2O5), a very satisfactory result with an increase of 75.26% when compared to the control treatment whose yield was 1,128 kg ha^{-1} . The regression equation model that best fitted the N doses was linear, inferring that the doses used were not sufficient to obtain maximum productivity. For the doses of P O_{25} , the quadratic model was used, with the maximum yield estimated at a dose of 85.56 kg ha^{-1} of P O_{25} (Figure 2).

$$Y = 24{,}66 + 0{,}0726*N - 0{,}0005*N^2 + 0{,}0671*P - 0{,}0005*P^2 \qquad R^2 = 0{,}70$$

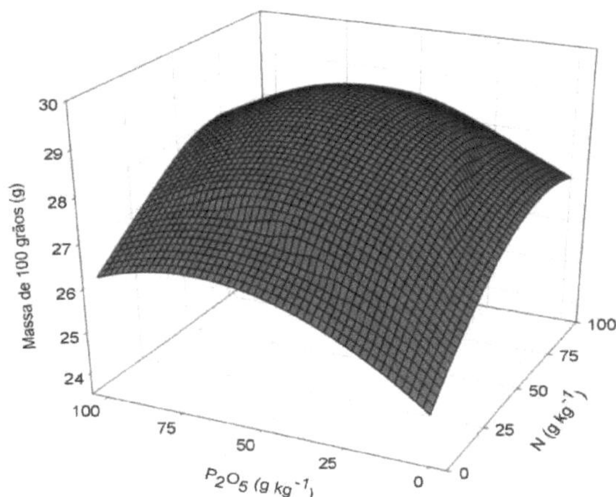

Figura 1 - Response surface for the mass of 100 grains as a function of doses of nitrogen and phosphorus applied to the soil.

The national average for dry grain production of cowpeas is 760 kg ha^{-1} (IBGE, 2005). Therefore, all the yields obtained, regardless of the treatments, exceeded this average, indicating that the caupi responded to the use of N and P. The yield achieved at the maximum dose (100 kg ha^{-1} of N and P O$_{25}$) was higher than that found by Oliveira et al. (2002), in the conditions of Areia-PB, which was 1,800 kg ha^{-1} using 300 kg ha^{-1} of P2O5 and lower than the result of Oliveira et al. (2004), who, when researching fava beans, obtained maximum yields of 2.7 kg ha^{-1} at a dose of 302 kg ha^{-1} of P2O5.

In terms of productivity, Santos et al. (2009) testing caupi varieties found that IPA-207 (1,187 kg ha^{-1}), IPA-206 (1,018 kg ha^{-1}), EPACE-10 (1,135 kg ha^{-1}), BRS Maratoã (1,016 kg ha^{-1}) and Costela de Vaca (1,191 kg ha^{-1}) had higher grain yields than the local materials: Cariri (611 kg ha^{-1}) and Canapu (685 kg ha^{-1}), which had statistically equal yields.

Table 5 - Grain yield, nitrogen and phosphorus content in the leaf, as a function of nitrogen and phosphorus doses applied to the soil

Dose of P O$_{25}$	N dose (kg ha)$^{-1}$					Average
(kg ha^{-1})	0	25	50	75	100	

		Grain yield (kg ha)$^{-1}$				
0	1.128	-	-	-	-	-
25	-	1.512	1.791	1.657	1.703	**1.665**
50	-	1.719	1.772	1.629	1.842	**1.740**
75	-	1.786	1.803	1.899	1.905	**1.848**
100	-	1.886	1.865	1.756	1.977	**1.871**
Average	-	**1.725**	**1.807**	**1.735**	**1.856**	**1.781**

ANAVA: CV (%): 10.4 F_{trat}: 4.56**

Regression: $Y = 1.226,34 + 1,83828*N + 12,3239**P - 0,0720198**P^2$ $R^2 = 0.81$

		Leaf nitrogen				
0	29,9	-	-	-	-	-
25	-	50,5	48,2	37,8	50,8	**46,8**
50	-	52,2	49,0	46,6	46,3	**48,5**
75	-	47,2	44,6	48,6	51,4	**47,9**
100	-	52,1	30,6	46,3	46,3	**43,8**
Average	-	**50,5**	**43,1**	**44,8**	**48,7**	**46,8**

ANAVA: CV (%): 7.6 F_{trat}: 15.13**

Regression: No model fitted the data

		Leaf		phosphorus		
		content (g kg)$^{-1}$				
0	3,9	-	-	-	-	-
25	-	3,0	3,9	2,8	4,0	**3,4**
50	-	3,9	3,0	3,1	3,4	**3,3**
75	-	3,1	3,3	3,3	3,6	**3,3**
100	-	3,0	2,6	2,9	2,5	**2,7**
Average	-	**3,2**	**3,2**	**3,0**	**3,3**	**3,2**

ANAVA: CV (%): 24.7 F_{trat}: 1.26^{ns}

Regression: No model fitted the data

___

**= P< 0.01; * = P<0.05;[ns] = not significant

$$Y = 1.226 + 1{,}8383*N + 12{,}3239**P - 0{,}07202**P^2 \qquad\qquad R^2 = 0{,}81$$

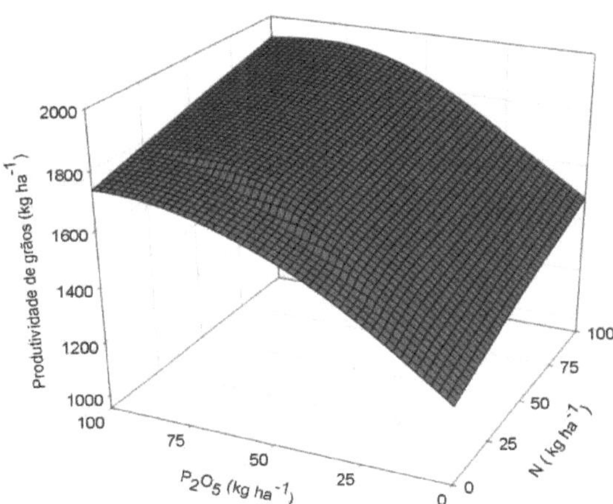

Figura 2 - Response surface for grain yield as a function of nitrogen and phosphorus doses applied to the soil.

Evaluating sources and doses of nitrogen and phosphorus in no-till beans, Valderrama et al. (2009) found that N sources, urea and coated urea, as well as P sources, triple superphosphate and coated triple superphosphate, did not differ in all the variables analyzed and grain yield, up to a dose of 120 kg ha^{-1} . Alvarez et al. (2005) and Meira et al. (2005) found that applying nitrogen fertilizer to bean plants increased grain yield, thus demonstrating that the soil is unable to meet the demands of the plants, which in turn have required high quantities of the element.

According to Table 5, the leaf nitrogen content showed a significant effect at 1% probability, but no model fitted the data. The leaf nitrogen content ranged from 29.9 g kg^{-1} (control) to 52.2 g kg^{-1} (25 kg ha^{-1} of N + 50 kg ha^{-1} of P2O5), an increase of approximately 74% with the lowest dose of N when compared to the control.

It can be seen that all the results obtained are within the range considered ideal for the bean crop. Malavolta et al. (1997) consider that the ideal N content in bean leaves ranges from 30 to 50 g kg^{-1} . This shows that the minimum dose of N applied to the soil was sufficient to raise the N content in the leaf to adequate levels. Thus, the critical level of N in the leaf diagnosed for cowpea for this experiment can be considered to be the overall average of the treatments, which was 46.8 g kg^{-1} .

25

These data disagree with those obtained by Soratto et al. (2002) who, when evaluating the effect of different doses and times of application of nitrogen as a top dressing on bean plants, found that doses of N applied as a top dressing did not increase the N content in the aerial part of the plants. However, they agree with the results of Andrade et al. (2002) who, studying the effect of increasing doses of N and P on the leaf content of macro and micronutrients, found that the increase in N fertilization, up to a dose of 120 kg ha^{-1} of N, increased the content of this nutrient in the leaves.

Table 5 also shows the results for leaf phosphorus content, where there was no significant effect for the treatments studied and no regression model fitted the observed data. It can also be seen that the zero dose treatment (control) had a higher average than the other treatments. According to Rosa et al., (2009); Maia et al., (2005) and Silva & Bohnen (2003) this can be explained by the dilutive effect of the nutrient on plant growth.

Thus, the overall average of 3.2 g kg^{-1} can be considered the critical level of P in the cowpea leaf. This value is slightly above the ideal range recommended by Malavolta et al. (1997), which varies from 2 to 3 g kg^{-1} , but it is below the sufficiency range recommended by MARTINEZ et al. (1999), which ranges from 4.0 to 7.0 g kg^{-1} .

Kikuti et al. (2006), when studying the effect of doses of N (0, 70, 140 and 210 kg ha^{-1}) and P2O5 (0, 100, 200 and 300 kg ha^{-1}) on the levels of macronutrients (N, P, K, Mg and Ca) in the aerial part of the bean plant, found no interaction in any of the characteristics studied.

4.3. Nitrogen and phosphorus content in the soil and crude protein

Table 6 shows the results of the nitrogen and phosphorus content in the soil and the crude protein content in the grains. It can be seen that the nitrogen content in the soil had no significant effect for the treatments studied and no regression model fitted the observed data.

Brito et al. (2011), evaluating five doses of N (2, 15, 30, 45 and 60 mg kg^{-1}) in the form of urea on the development of beans (cultivar Carioca) and caupi (cultivar CNC x 284-4E,) found no influence on soil N as a result of the nitrogen doses applied to a medium-textured, dystrophic red-yellow latosol in Piracicaba (SP). This behavior is in the same order as that presented in this study.

The N requirement for the bean plant in vârzeas was higher than in the irrigated upland cultivation system, where the maximum responses occurred with 72 kg ha^{-1} (SILVEIRA & DAMASCENO, 1993), 109 kg ha^{-1} (SILVA & SILVEIRA, 2000), 137 kg ha^{-1} (BARBOSA FILHO & SILVA, 1994; STONE &MOREIRA, 2001) and 150 kg ha^{-1} (CARDOSO ET AL.1978)

For the phosphorus content in the soil, there was a significant effect at the 1% probability level for the treatments studied, which ranged from 18.8 mg dm^{-3} (control) to 70.0 mg dm^{-3} (100 kg ha^{-1} of N + 100 kg ha^{-1} of P2O5). However, the regression equation that best followed the data was positive and linear for the doses of N and P2O5 studied. The doses of P2O5 showed a linear effect of great magnitude, indicating that the phosphorus content in the soil increased with the doses of P2O5 applied. This behavior can be understood by the fact that phosphate fertilization has a residual effect on the soil (Table 6, Figure 3).

The data partially correspond to those obtained by Silva (2007), who applied doses of phosphorus to cowpea cultivation in three successive crops and found an increase in the $P O_{25}$ content in the soil of a Neossolo Regolitico with a sandy texture, in the municipality of Areia-PB.

Thus, the critical levels of N and P in the soil for this experiment will be considered to be the respective averages of the treatments, with 0.96 g kg^{-1} being the critical level of N in the soil and 48.2 mg dm^{-3} being the critical level of P in the soil.

Table 6 - Nitrogen and phosphorus content in the soil and crude protein, as a function of nitrogen and phosphorus doses applied to the soil

Dose of P2O$_5$ (kg ha^{-1})	N dose (kg ha)$^{-1}$ 0	25	50	75	100	Average
		Nitrogen				
0	0,93	-	-	-	-	-
25	-	1,00	0,89	0,93	1,00	**0,95**
50	-	1,03	0,96	0,98	1,00	**0,99**
75	-	0,93	1,00	0,85	0,95	**0,93**
100	-	1,12	0,91	0,93	0,97	**0,98**
Average	-	**1,02**	**0,94**	**0,92**	**0,98**	**0,96**
ANAVA:		CV (%): 12.9			F$_{trat}$: 0.99^{ns}	
Regression:	No model fitted the data					
		Phosphorus content in the soil (mg dm)$^{-3}$				
0	18,8	-	-	-	-	-
25	-	32,0	33,8	43,8	41,5	**37,7**
50	-	28,0	35,5	45,0	66,0	**43,6**
75	-	40,3	42,7	56,0	69,5	**52,1**

100	-	59,3	41,3	67,7	70,0	**59,5**
Average	-	**39,9**	**38,3**	**53,1**	**61,7**	**48,2**
ANAVA:		CV (%): 40.1			Ftrat: 2.79**	
Regression:	Y = 13.38 + 0.2944**N + 0.2691**P					$R^2 = 0.81$

Crude protein content (dag kg)$^{-1}$ ----------

0	18,0	-	-	-	-	-
25	-	19,9	19,3	19,8	20,6	**19,9**
50	-	20,4	20,2	19,5	20,9	**20,3**
75	-	20,4	21,5	20,1	19,2	**20,3**
100	-	21,2	20,6	22,7	21,0	**21,4**
Average	-	**20,5**	**20,4**	**20,5**	**20,4**	**20,5**
ANAVA:		CV (%): 10.1			Ftrat: 1.07^{ns}	
Regression:	No model fitted the data					

**= P< 0.01; * = P<0.05;^{ns} = not significant

$$Y = 13,38 + 0,2944**N + 0,2691**P \qquad R^2 = 0,81$$

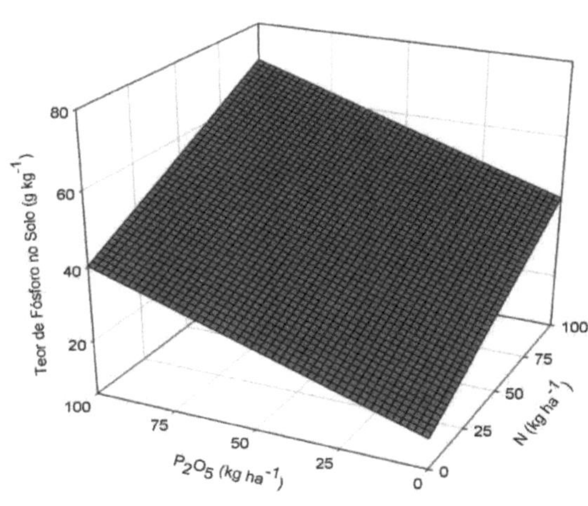

Figure 3 - Response surface for soil phosphorus content as a function of nitrogen and phosphorus doses applied to the soil.

The residual effect of phosphate fertilization can be understood as the amount of total phosphorus applied to the soil that tends to remain available to plants over a period of time. Phosphorus tends to accumulate in the soil, so that the effect of corrective fertilization lasts for several years. On the other hand, the phosphorus content in the soil can also be gradually increased with annual applications (FREIRE et al., 1998).

As for crude protein (Table 6), there was no significant effect for the treatments studied and no regression model fitted the observed data. Even though there was no statistically significant effect, it can be seen that the average of 22.78 % (75 kg ha^{-1} of N + 100 kg ha^{-1} of P2O5) obtained an increase of 26.5% in crude protein content when compared to the 18.04% treatment (control). This result can be considered satisfactory.

Ramos Junior et al. (2005), evaluating the behavior of fifteen bean cultivars in Botucatu (SP), found an average crude protein content of 20.5% in the grains. These data corroborate the general average found in this work, being lower than the 22.7% and 25.4% values found respectively by Pàrraga et al. (1981) and Pimentel et al. (1988) in 200 cultivars and 20 strains of beans.

Farinelli et al. (2006) also found an increase in crude protein content when using doses of 0, 40, 80, 120, 160 kg ha^{-1} of N in a dystrophic red nitossolo.

The composition of cowpeas varies depending on the cultivar, as Castellón et al. (2003), analyzing six cowpea cultivars (Br14, Br 9, Br 17, CNC 0434, Vita 7 and CE 315), found increases in crude protein ranging from 21.6 to 24.7%.

4.4. Economic analysis of fertilization

Table 7 shows the grain yield values estimated by the regression model (Figure 2), gross revenue, expenditure on nitrogen and phosphate fertilizers and net revenue as a function of the doses of N and P O_{25} applied to the soil. According to the results, the highest net revenue was R$6,191.53, corresponding to a maximum physical efficiency yield of 1,938 kg ha^{-1} with the doses of 100 kg ha^{-1} of N and 86 kg ha^{-1} of P O_{25} applied to the soil. To obtain this net revenue, the producer would have had to invest R$262.01 in nitrogen and phosphate fertilizers, obtaining a net revenue profit of R$2,108.95, representing an increase of approximately 50% when compared to the treatment without N and P fertilization O_{25}.

The main fertilizer tables for beans in use in the country recommend 30 to 40 kg ha^{-1} of N and 50 to 90 kg ha^{-1} of P $O_{25,}$ expected productivity of 1,2000 to 2,500kg ha^{-1} for the state of Minas Gerais (CFSMG, 1999). Freire Filho (2005) suggests a chemical fertilizer recommendation for caupi of 20 kg ha^{-1} of N in top dressing and 20 to 60 of P2O5. The State Soil Fertility Commission of

Table 7 - Estimated caupi bean production, gross revenue, expenditure on nitrogen and phosphate fertilizers and net revenue as a function of doses of N and P2O5 applied to the soil

Dose of N	Dose of P2O5	Estimated productivity	Gross Revenue	Fertilizer costs	Revenue Liquid
kg ha^{-1}	kg ha^{-1}	kg ha^{-1}	R$	R$	R$
0	0	1.226	4.082,58	0,00	4.082,58
25	25	1.535	5.111,55	71,75	5.039,80
25	50	1.708	5.687,64	115,00	5.572,64
25	75	1.792	5.967,36	158,25	5.809,11
25	86	1.800	5.994,00	176,51	5.817,49
25	100	1.785	5.944,05	201,50	5.742,55
50	25	1.581	5.264,73	100,25	5.164,48
50	50	1.754	5.840,82	143,50	5.697,32
50	75	1.838	6.120,54	186,75	5.933,79
50	86	1.846	6.147,18	205,01	5.942,17
50	100	1.831	6.097,23	230,00	5.867,23
75	25	1.627	5.417,91	128,75	5.289,16
75	50	1.800	5.997,33	172,00	5.825,33
75	75	1.883	6.270,39	215,25	6.055,14
75	86	1.892	6.300,36	233,51	6.066,85
75	100	1.877	6.250,41	258,50	5.991,91
100	25	1.673	5.571,09	157,25	5.413,84
100	50	1.846	6.147,18	200,50	5.946,66
100	75	1.929	6.423,57	243,75	6.179,82
100	86	1.938	6.453,54	262,01	6.191,53
100	100	1.923	6.403,59	287,00	6.116,59

Pernambuco (2008) suggests for cowpeas 20 kg ha^{-1} of N at planting and 30 kg ha^{-1} of N in top dressing and 20 to 60 kg ha^{-1} of P2O5, for an expected yield of 1,800 kg ha^{-1} . The Federal University of Ceará (1993) recommends 20 kg ha^{-1} of N at planting and 30 to 110 kg ha^{-1} of P2O5, for an expected yield of between 1,000 and 1,200 kg ha $.^{-1}$

Although the treatments 100 kg ha^{-1} of N + 86 kg ha^{-1} P2O5, 100 kg ha^{-1} of N + 75 kg ha^{-1} of P2O5 and 100 kg ha^{-1} of N + 100 kg ha^{-1} of P2O5 showed higher net revenue values (Table 7), it is advisable to use the doses of 50 kg ha^{-1} of N and 86 kg ha^{-1} P2O5 to obtain productivity with a greater margin of safety, a reduction in costs and maximum economic and environmental efficiency.

Silva et al. (2004) found that grain yield was influenced by nitrogen doses and significant increases were obtained with applications of 75 to 100 kg ha^{-1} of N. Testing phosphorus doses and application methods, Silva et al. (2010) concluded that the dose of 90 kg ha^{-1} of P $_{25}$ provided the highest grain yield for cowpeas grown on a Yellow Latosol.

The highest net income (R$6,191.53) was estimated for a grain yield of 1,938 kg ha^{-1} , corresponding to the application of 100 kg ha^{-1} of N + 86 kg ha^{-1} of P O$_{25}$. However, applying just 25 kg ha^{-1} of N + 75 kg ha^{-1} of P O$_{25}$ would produce 1,792 kg ha^{-1} of grain, with a net income of R$5,809.11, which corresponds to 94% of the estimated maximum net income.

5. FINAL CONSIDERATIONS

Increasing the doses of nitrogen and phosphorus applied to the soil for the cowpea crop increased the dry matter of the aerial part, the mass of 100 grains, grain yield and the levels of nitrogen and phosphorus in the leaf.

The critical levels of nitrogen and phosphorus in the diagnostic leaf were 46.8 and 3.24 g kg^{-1} , respectively. In the soil, the critical level of phosphorus extracted with the Mehlich extractor^{-1} was 48.2 mg dm^{-3} .

Applying just 25 kg ha^{-1} of N + 75 kg ha^{-1} of P $_{25}$ would produce 1,792 kg ha^{-1} of grain, with a net income of R\$ 5,809.11, which is 94% of the maximum estimated net income.

6. BIBLIOGRAPHICAL REFERENCES

AL-KAISI, M & LICHT, M.A. Effect of strip tillage on corn nitrogen uptake and residual soil nitrate accumulation compared with no-tillage and chisel plow. **Agronomy Journal**, v.96, n. 4, p. 1164-1171, 2004.

ALMEIDA, A. A. F.; LOPES, N. F.; OLIVA, M. A. Development and partitioning of assimilates in *Phaseolus vulgaris* submitted to three doses of nitrogen and three levels of light. **Pesquisa Agropecuària Brasileira**, Brasilia, v.23, n.8, p.837-847, aug. 1988.

ALVAREZ, V. V. H. Correlation and calibration of soil analysis methods. In: Alvarez V., V. H.; Fontes, L. E. F. & Fontes, M. P. F. (Eds.). Soil in the major morphoclimatic domains of Brazil and sustainable development. **Viçosa**: SBCS/UFV/DPS, 2005. p.615-646.

ALVES, A. C. **Methods for quantifying N-NH volatilization$_3$ in soil fertilized with urea.** 2006. 41p. Dissertation (Master's Degree in Animal Science) - School of Animal Science and Food Engineering, University of São Paulo, Pirassununga, 2006.

AMADO, T. J. C.; MIELNICZUK, J.; AITA, C. Nitrogen fertilization recommendation for corn in RS and SC adapted to the use of soil cover crops, **under no-tillage. Revista Brasileira de Ciências do Solo,** Viçosa, v. 2, n. 1, p. 241 - 248, 2002.

ANDRADE JUNIOR, A. S. Viability **of irrigation, under climatic and economic risk, in the micro-regions of Teresina and Litoral Piauiense**. Piracicaba: ESALQ, 2000. 56p. Thesis (Doctorate).

ANDRADE, M. J. B. *et al.* Foliar contents of macro and micronutrients in bean plants (CVs Carioca and Pérola) as a function of nitrogen and phosphorus doses. *In:* CONGRESSO NACIONAL DE PESQUISA DE FEIJAO,7, 2002, Viçosa - MG. *Expanded Abstracts.* **Viçosa**: UFV, 2002. p.765-767

ANDRADE, C. A. B.; PATRONI, S. M. S.; CLEMENTE, E.; SCAPIM, C.A. Productivity and nutritional quality of beans under different fertilizations, 2004. Available at: <http://**www.editora.ufla.br/revista/28_5/art15.pdf**> **Accessed on April 18, 2005.**

ANDRADE, C. A. de B. **Fertility limitations and the effect of limestone for the bean plant (*Phaseolus vulgaris* L.) in varzea soils in the south of Minas Gerais**. Lavras: UFLA, 1997. 107 p. (Thesis - Doctorate in Plant Science).

AOAC. Official Methods of Analysis of the Association of Official Analytical Chemists. 16. ed. **Washington**: DC, Chap.32, p.25-28, 1995.

ARAUJO, E. S.; MEDEIROS, A. F. A.; DIAS, F. C.; URQUIAGA, S.; BODDEY, R. M. & ALVES, B. J. R. Quantification of soil N derived from soybean roots using the isotope[15] **N. Revista Universidade Rural**, Seropédica, RJ v. 24, n.1, p. 7-12, 2004.

ARAUJO, L. A. N de. FERREIRA, M.E & CRUZ, M. C. P DA. Nitrogen fertilization in corn cultivation. **Pesquisa Agropecuària Brasileira,** Brasilia, v.39, n.8, p.771-777, 2004.

AZEVEDO, W. R.; FAQUIN, V.; FERNANDES, L. A.; OLIVEIRA JÙNIOR, A. C. Phosphorus availability for flooded rice under the residual effect of limestone, gypsum and farmyard manure applied to the bean crop. **Revista Brasileira de Ciência do Solo**, Viçosa, v, 28, p. 995-1004, 2004.

BARBOSA FILHO, M. P.; FAGERIA, N. K.; SILVA, O. F. Application of nitrogen to irrigated bean crops. **EMBRAPA**, 2001, 8 p. (Technical circular, 49).

BARBOSA FILHO, M.P. & SILVA, O. F. DA. Fertilization and liming for beans in cerrado soil. **Pesquisa Agropecuària Brasileira**, Brasilia, v. 35, n.7, p. 1317-1324, 2000.

BASSAN, D. A. Z.; ARF, O.; BUZETTI, S.; CARVALHO, M. A. C.; SANTOS, N. C .B. & SA, M .E. Seed inoculation and application of nitrogen and molybdenum in winter bean crops: Production and physiological quality of seeds. **Revista Brasileira Sementes,** v. 23, p. 76 - 83, 2001.

BRAZIL, Ministry of Agriculture. Research and Experimentation Office. Pedology and soil fertility team. I. Exploratory and reconnaissance survey of soils in the state of Paraiba. II. Interpretation for the agricultural use of the soils of the state of Paraiba. **Rio de Janeiro**, 1972. 683 p. (Boletim tècnico, 15; SUDENE. Série Pedologia, 8)

BRAZIL. Ministry of Agriculture, Livestock and Supply. **Rules for Seed Analysis**. Ministry of Agriculture, Livestock and Supply. Secretariat for Agricultural Defense. Brasilia: MAPA/ACS, 2009. 395p.

BRITO. M. DE M. P.; MURAOKA, T.; SILVA, E. C. DA. Contribution of biological nitrogen fixation, nitrogen fertilizer and soil nitrogen to the development of beans and cowpeas. **Bragantia,** Campinas, v. 70, n. 1, p.206-215, 2011.

BRITO. M. DE M. P.; MURAOKA, T.; SILVA, E. C. DA. Soil nitrogen uptake, fertilizer and symbiotic fixation in cowpea *(Vigna unguiculata (L) WALF)* and common bean (*Phaseolus vulgaris L.*) determined using[15] **N. Revista Brasileira de Ciência do solo**, v. 3, p. 895 - 905. 2009.

CARTWRIGHT, B.; TILLER, K. G.; ZARCINAS, B. A.; SPOUNCER, L. R. The chemical assessment of the baron status of soils. **Australian Journal of Soil Research.** Victoria, v. 21, p. 321 - 332, 1983.

CARVALHO, A. M. DE.; FAGERIA, N. K.; OLIVEIRA, I. P. DE. DE.; KINJO, T. Bean plant response to phosphorus application in cerrado soils. **Revista Brasileira de Ciência do Solo**, n. 19, p. 61-67. 1995.

CARVALHO, M. A. C. Influence of nitrogen sources and application methods on the physiological quality of winter bean seeds (*Phaseolus vulgaris L.*). **Informativo ABRATES**, Londrina, v.9, n.1/2, p.118, 1999.

CARVALHO, M. A. C.; FURLANI JUNIOR, E.; ARF, O.; SA, M. E. DE; PAULINO, H. B.; BUZETTI, S. Doses and times of nitrogen application and foliar levels of this nutrient and chlorophyll in bean. **Revista Brasileira de Ciência do Solo**, v.27, p.445-450, 2003.

CARVALHO, M.A.C.; ARF, O.; SA, M.E.; BUZETTI, S.; SANTOS, N.C.B.; BASSAN, D.A.Z. Productivity and quality of bean seeds (*Phaseolus vulgaris L.*) under the influence of nitrogen installments and sources. **Revista Brasileira de Ciência do Solo**, Viçosa, v. 25, n.3, p. 617-624, 2001.

CASTELLÓN, R. E. R; ARAÙJO, F. M. M. C.; RAMOS, M. V.; ANDRADE NETO, M.; FREIRE FILHO, F. R.; GRANGEIRO, T. B.; CAVADA, B. S. Elemental composition and characterization of the lipid fraction of six caupi cultivars. **Revista Brasileira de Engenharia Agricola e Ambiental**, Campina Grande, PB, v.7, n.1, p.149-153, 2003.

CERETTA, C. A.; BASSO, C. J.; HERBES, M. G.; POLETTO, N. & SILVEIRA, M. J. Production and decomposition of phytomass of winter soil cover plants and corn, under different nitrogen fertilizer managements. **Ciência Rural**, v. 32, p. 49-54, 2002.

SOIL FERTILITY COMMISSION OF THE STATE OF MINAS GERAIS - CFSEMG. Recommendations for the use of correctives and fertilizers in Minas Gerais: 5ª Approximation. RIBEIRO, A.C.; GUIMARAES, P. T. G.; ALVAREZ, V. (Ed.). Viçosa: **UFV**, 1999. 259 p.

STATE SOIL FERTILITY COMMISSION, Fertilization recommendations for the state of Pernambuco, 2ª approach, **Pernambuco**, 2008.

CoNAB - National Supply Company. Monitoring the Brazilian harvest: grains, sixth survey, March 2011. **CONAB**, 2012.

CRAVO, M. S.; SMYTH, T. J. High input cropping systems in the Brazilian Amazon. In: Smyth, T. J.; Raun, W. R.; Bertsch, F. Manejo de suelos tropicales en Latino america. Talles Latinoamericano de Manejo de Suelos Tropicales 2. San José, 1990. **North Carolina States University**, 1991, p. 145-156.

CRUSCIOL C. A. C.; SORATTO, R. P.; DA SILVA L. M.; LEMOS. L. B. Sources and doses of nitrogen for bean in succession to grasses in the no-till system. **Revista Brasileira de Ciência do**

Solo, v. 31, p. 545-552, 2007.

CRUSCIOL, C. A. C.; LIMA, E. V.; ANDREOTTI, M.; NAKAGAWA, J.; LEMOS, L. B.; MARUBAYASHI, O. N. Effect of nitrogen on the physiological quality, productivity and characteristics of bean seeds. **Revista Brasileira de Sementes,** Brasilia, v. 25, n. 1, p. 108-115, 2003.

DUTRA, A. S.; BEZERRA, F. T. C.; NASCIMENTO, P. R.; LIMA, D. DE C. Productivity and physiological quality of cowpea seeds as a function of nitrogen fertilization. **Rev. Ciênc. Agron.,** v. 43, n. 4, p. 816-821, Oct-Dec, 2012.

EMBRAPA - Brazilian Agricultural Research Corporation. Brazilian Soil Classification System. Brasilia: Embrapa Produçào **de** Informaçào; **Rio de Janeiro**: Embrapa Solos, 1999. 412 p.

EMBRAPA. Manual of soil analysis methods. **Rio de Janeiro,** Embrapa Solos, 1997. 212p.

FAGERIA, N. K.; BALIGAR, V. C. Response of lowland rice and common bean grown in relation to soil fertility levels on a vàrzea soil. **Fertilizer Research,** Dordrecht, v.45, n.8, p.13-20, 1996.

FAGERIA, N. K.; SANTOS, A. B. Phosphate fertilization for the bean plant in vàrzea soil. **Revista Brasileira de Engenharia Agricola Ambiental**, Campina Grande, v.2, n.2, p.124-127, 1998.

FAGERIA, N.K. Calibration of phosphorus analysis for rice in greenhouse conditions. **Pesquisa Agropecuària Brasileira**, Brasilia, v.25, n.4, p.579-586, 1990.

FAQUIN, V.; ANDRADE, C. A. B.; FURTINI NETO, A. E.; ANDRADE, A. T.; CURI, N. Response of the bean plant to the application of limestone in varzea soils of southern Minas Gerais. **Revista Brasileira de Ciência do Solo,** Campinas, v.22, n.4, p.651660, Oct./Dec. 1998.

FARINELLI, R.; LEMOS, L. B.; PENARIOL, F. G.; EGÉA, M. M.; GASPAROTO, M. G. Nitrogen fertilization of cover crops in no-till and conventional tillage. **Pesquisa Agropecuària Brasileira,** Brasilia, v.41, n.2, p.307-312, feb. 2006.

FILGUEIRA, F. A. R. Novo Manual de Olericultura: Agrotecnologia moderna na produção e comercializaçao de hortaliças, **Viçosa,** 2000, 402 p.

FREIRE FILHO, F. R. **Genetics of caupi**. In: ARAÙJO, J.P.P.; WATT, E.E. prg. Cowpeas in Brazil. Brasilia: IITA/EMBRAPA, 1988. p. 159 - 229.

FREIRE FILHO, F. R. **Feijâo-caupi Avanços tecnológicos**. Brasilia: Embrapa Informaçào Tecnològica; 2005. 519 p. (Embrapa Technological Information).

FREIRE, F. M.; OLIVEIRA, L, A.; FRANÇA, G. E. COUTO, L.; ALVES ,U. M. C. Effect of phosphorus-water ratio on maize nutrition in Quartz Sand. In: CONGRESSO NACIONAL DE

MILHO E SORGO, 22. Recife, 1998. **Abstracts.** Recife: IPA/ EMBRAPA 1998, p.143.

HINSINGER, P. Biology availability of soil inorganic P in the rhizosphere as affected by root-induced chemical changes: a review. **Plant and Soil**, 237:173-195, 2001.

HORST, W. J.; KAMH, M; JIBRIN, J. M.; CHUDE, V.O. Agronomic measurements for increasing P availability to crops. **Plant and Soil**, 237:211-223, 2001.

IBGE - Brazilian Institute of Geography and Statistics. **Systematic survey of agricultural production**: monthly forecast survey and monitoring of agricultural harvests in the calendar year. IBGE, v. 23, n. 09, p. 1-80, 2010.

IBGE. Brazilian Institute of Geography and Statistics. IBGE Automatic Retrieval System - SIDRA, 2005. Available at: http://www.sidra.ibge.gov.br/. Accessed on: October 2009.

KELLER, G. D.; MENGEL, D. E. Ammonia volatilization from nitrogen fertilizers surface applied to no-till corn. **Soil Science Society of America Journal**, Madison, v.50, p.1060-1063, 1986.

KIKUTI1, H.; ANDRADE, M. J. B. DE.; CARVALHO, J. G. DE.; MORAIS, A. R. DE. Macronutrient content in the aerial part of bean plants as a function of nitrogen and phosphorus doses. **Bragantia**, Campinas, v.65, n.2, p.347-354, 2006.

KORNDORFER, G. H., LARA-CABEZAS, W. A., HOROWITZ, N. Agronomic efficiency of reactive natural phosphates in corn cultivation. **Science agricultural**, v.56, n.2, p.391-396, 1999.

LARA, W. A. R.; TRIVELIN, P. C. O. Efficiency of a static semi-open collector in the quantification of volatilized NNH3 for urea applied to soil. **Revista Brasileira de Ciência do Solo**, Campinas, v. 14, p. 345-352, 1990.

LYNCH, J. P.; BROWN, K. M. tropsoil foraging: na architectreral adaptation to low Phosphorus availability. **Plant and soil**, v. 3, n.2, p. 225 - 237, 2001.

MAIA, C. E.; CANTARUTTI, R. B. Accumulation of nitrogen and carbon in the soil by continuous organic and mineral fertilization of corn. **Revista Brasileira de Engenharia Agricola e Ambiental**, v.8, n.1, p. 39-44, 2004.

MAIA, C. E.; MORAIS, E. R. C. DE.; PORTO FILHO, F. de Q.; GUEYI, H. R.; MEDEIROS, J. F. de. Foliar nutrient content in melon irrigated with water of different salinities. **Revista Brasileira de Engenharia Agricola e Ambiental,** v.9, p.292-295, 2005.

MALAVOLTA, E. ABC da adubaçào. **Sào Paulo**: Ceres, 1989. 250p.

MALAVOLTA, E. Elementos de nutriçào mineral de plantas. Sào Paulo: Editora Agronômica, **Revista Ceres,** 1980.

MALAVOLTA, E.; VITTI, G. C.; OLIVEIRA, S. C. Avaliaçào do estado nutricional das plantas: principios e aplicações. 2.ed, **Piracicaba**: Potafos, 1997. 319p.

MALAVOLTA, E.; VITTI, G. C.; OLIVEIRA, S.A. Avaliaçào do estado nutricional das plantas, principios e aplicações. 2.ed.**Piracicaba**: POTAFOS, 1997. 319 p.

MARSCHENER, H. Mineral nutrition of higher plants, San Diego. **Academic Prese**, 1995, 889 p.

MARSCHNER, H. Mineral nutrition of higher plants. Orlando: **Academic**, 2002. 889p.

MARTENS, D. C.; WESTERMANN. D. T. Fertilizer applications for correcting micronutrient deficiencies. In: MORTVEDT, J. J. et al. (ED). Micronutrients in agriculture. 2nd ed. Madison: **Soil Science Society of America**, 1991. p. 549-591.

MARTINEZ, H. E. P. Zn translocation as a function of doses applied to bean and coffee plants via the root system. **Ciência** Rural, Santa Maria, v. 35, n. 3, p. 491 - 497, 2005.

MARTINEZ, H. E. P.; CARVALHO, J. G.; SOUZA, R. B. Foliar diagnosis. In: RIBEIRO, A.C.; GUIMARAES, P. T. G.; ALVAREZ V, V. H. (Eds). Recommendations for the use of soil improvers and fertilizers in Minas Gerais. 5th Approximation. **Viçosa**: Fertility Commission of the State of Minas Gerais - CFSEMG, 1999. p.143-168.

MEIRA, F. A.; SA, M. E.; BUZETTI, S.; ARF, O. Doses and times of nitrogen application to irrigated no-till bean crops. **Pesquisa Agropecuâria Brasileira**, Brasilia, v.40, n.4, p.383-388, 2005.

MENGEL, K.; KIRKBY, A. Principles of plant nutrition. Bern: **International Potash institute**, 1987, 687p.

MIRANDA, L. N.; AZEVEDO, J. A.; MIRANDA, J. C. C.; GOMES, A.C. Bean productivity in response to phosphate fertilization and irrigation regimes in cerrado soil. **Pesquisa Agropecuària Brasileira**, Brasilia, v.35, n.4, p.703710, 2000.

MIRANDA, P.; COSTA, A. F.; OLIVEIRA, L. R.; TAVARES, J. A.; PIMENTEL, M. L.; LINS, G. M. L. Behavior of cultivars of *Vigna unguiculata* L) Walp, in single and intercropped systems. IV - erect and semi-erect type. **Pesquisa Agropecuària Pernambucana**, Recife, v. 9, n. especial, p. 95-105, 1996.

MOREIRA, A.; MALAVOLTA, E.; MORAES, L. A. C. Efficiency of phosphorus sources in alfalfa centroena grown in Yellow Latosol. **Pesquisa Agropecuària Brasileira**, Brasilia, v.37, n.10, p.1459-1466, 2002.

MOREIRA, A.; MALAVOLTA, E.; MORAES, L. A. C. Efficiency of phosphorus sources in alfalfa centroena grown in Yellow Latosol. **Pesquisa Agropecuària Brasileira,** Brasilia, v.37, n.10, p.1459-1466, 2002.

MOURA, W. M.; LIMA, P. C.; CASALI, V. W. D.; PEREIRA, P. R. G. Nutritional efficiency for phosphorus in pepper lines. **Horticultura Brasileira,** Brasilia, v. 19, n. 3, p. 306-312, 2001.

NABLE, R. O.; BANUELOS, G. S.; PAULL, J. G. Baron toxicity. **Plant and soil**, the Hague, v. 193, n ½, p. 181 - 198, June 1997.

NASCIMENTO, M. S.; ARF, O.; SILVA, M. G. Bean plant response to nitrogen topdressing and molybdenum foliar application. **Acta Scientiarum**. Agronomy. Maringà, v. 26, n° 2, p. 153 - 159, 2004.

OLIVEIRA, A. P. SILVA, V. R. F; ARRUDA, F. P. de; NASCIMENTO, I. S. do; ALVES, A. U. Yield of cowpea as a function of doses and forms of nitrogen application. **Horticultura Brasileira**, v.21, n1. Brasilia, 2003.

OLIVEIRA, A. P.; ALVES, E. U.; ALVES, A. U.; DORNELAS, C. S. M.; SILVA, J. A.; PÔRTO, M. L. Fava bean production as a function of phosphorus doses. **Horticultura Brasileira**, Brasilia, v.22, n.3, p.543-546, 2004.

OLIVEIRA, A. P.; ARAÙJO, J. S.; ALVES, E. U.; NORONHA, M. A. S.;CASSIMIRO, C. M.; MENDONÇA, F. G. Yield of cowpea grown with cattle manure and mineral fertilizer. **Horticultura Brasileira**, Brasilia, v. 19, n. 1, p. 81-84, 2001.

OLIVEIRA, F. H. T; ARRUDA, J. A.; SILVA, I. F.; ALVES, J. C. Sampling for soil fertility evaluation as a function of the sampling instrument and types of soil preparation. **Revista Brasileira de Ciência do Solo**, 31:973-983, 2007.

OLIVEIRA, I. P. & CARVALHO, A. M. The caupi crop in the climate and soil conditions of the humid tropics of semi-arid Brazil. In: ARAÙJO, J. P. P. & WATT, E. E. O caupi no Brasil. **Brasilia**: EMBRAPA-CNPAF, 1988. p.65-69.

OLIVEIRA, I. P.; ARAUJO, R. S.; DUTRA, L. G. Mineral nutrition and biological nitrogen fixation. In: ARAUJO, R.S.; RAVA, C.A.; STONE, L.F.; ZIMMERMANN, M.J.O. (Coord.). **Common bean crop in Brazil**. Piracicaba: Potafos, 1996. p. 169- 221.

PARISCHA, N. S.; FOX, R. L. Plant nutrient sulphur in the tropic sand subtropics. **Advances in Agronomy**, New York, v. 50, p. 209 - 269, 1993.

PÂRRAGA, M.S.; JUNQUEIRA NETTO, A.; PEREIRA, P.; BUENO, L.C.S.; PENONI, J.S. Evaluation of the total protein content of two hundred bean cultivars (Phaseolus vulgaris L.) for genetic improvement. **Ciência e Pràtica**, Lavras, v.5, n.1, p.07-17, 1981.

PARRY. M. M. et al. Macronutrients in caupi grown under two doses of phosphorus at different planting times. **Revista Brasileira de Engenharia Agricola e Ambiental** v.12, n.3, p. 236-242. Campina Grande: UAEAg/UFCG, 2008.

PASTORINI, L. H.; BACARIN, M. A.; LOPES, N. F.; LIMA, M. G. S. Initial growth of bean submitted to different doses of phosphorus in nutrient solution. **Revista Ceres**, Viçosa, v.47, n.270, p.219-228, 2000.

PAULINO, H. B. **Parcellation of two sources in cover crops and via fertigation and cost of production in bean cultivation.** Ilha Solteira. 1998. 84p. Dissertation (Master's Degree in Agronomy) - Faculty of Engineering of Ilha Solteira, Paulista State University, Ilha Solteira, 1998.

PIMENTEL, M.L.; MIRANDA, P.; COSTA, A. F.; MIRANDA, A. B. Nutritional study of common bean strains (*Phaseolu svulgaris* L.). **Revista Brasileira de Sementes**, Brasilia, v.10, n.2, p.55-65, 1988.

PÔRTO, M. L.; OLIVEIRA, A.P.; ALVES, J. C.; BRUNO, G. B.; ALVES, E. U.; SANTOS, E. O. Yield of cowpea as a function of the residual effect of phosphorus. **Horticultura Brasileira**, Brasilia, v.23, n.2, p.34 - 44, 2005.

POTAFOS. *Brazil*: apparent consumption of fertilizers. Available at: <www.potafos.org>. Accessed on: August 8, 2006.

RAIJ, B. V,; CANTARELLA, H,; QUAGGIO, J.A,; FURLANI, A. N. C. recomendaçoes e adubaçào e calagem para o estado de Sào Paulo, **Campinas**: Instituto Agronômico & Fundaçào IAC, 1996. 285 p. (Boletim técnico 100)

RAIJ, B. V. **Soil fertility and fertilization.** Piracicaba: Ceres, Potafos, 1991.

RAIJ, B. van. Evaluation of soil fertility. Piracicaba, Potash & Phosphate Institute and International Potash Institute, **São Paulo**, 1981. 142p.

RAMOS JUNIOR, E. U.; LEMOS, L. B. SILVA, T. R. B. DA.Production components, grain yield and technological characteristics of bean cultivars. **Bragantia**, Campinas, v.64, n.1, p.75-82, 2005.

RAPASSI, R. M. A.; SA, M. E. DE.; TARSITANO,M. A. A.; CARVALHO, M. A. C. DE.; PROENÇA, E . R.; NEVES, C. M. T. DE C.; COLOMBO, E. C. M. Comparative economic analysis after one year of irrigated bean cultivation, in winter, in conventional and no-till systems, with different sources and doses of nitrogen. **Bragantia**, Campinas, v.62, n.3, p.397-404, 2003.

RIBEIRO JUNIOR, J. I. Statistical analysis in SAEG. **Viçosa**: Federal University of Viçosa, 2001. 301 p.

ROSA, C. M.; CASTILHOS, R. M. V.; VAHL, L. C.; CASTILHOS, D. D.; PINTO, L. F. S.; OLIVEIRA, E. S.; LEAL, O. A. Effect of humic substances on potassium absorption kinetics, plant growth and nutrient concentration in *Phaseolus vulgaris L.* **Revista Brasileira de Ciência do Solo,** v.33, n.4, 2009.

ROSOLEM, C. A.; MARUBAYASHI, O. M. Be a doctor of your bean plant. Informaçoes agronômicas, **Piracicaba**, v. 68, p. 1 - 16, 1994.

SANTANA, G. S.; FERNANDES, F. M.; ANDRADE, J. A. C.; FERNANDES DEUS,A. C. Response of maize to doses of NPK, in two years of cultivation, in Latosol with corrective phosphate fertilization. In: BRAZILIAN CONGRESS OF SOIL SCIENCE, 31st, 2007, Gramado. Proceedings... **Gramado**, 2007.

SANTOS, A.B; FAGERIA, N. K.; SILVA, O. F.; MELO, M. L. B. Bean response to nitrogen management in tropical varzeas. **Pesquisa Agropecuâria Brasileira,** Brasilia, v.38, n.11, p.1265-1271, 2003.

SANTOS, J. F. DOS.; GRANGEIRO, J. I. T.; BRITO, L. M. P.; OLIVEIRA, M. M. DE.; OLIVEIRA, M. E. C. DE. New varieties of caupi for the Brejo Paraibano micro-region. **Tecnologia & Ciência Agropecuâria** Joào Pessoa, v.3, n.3, p.07-12, sep. 2009.

SANTOS, J. P. V. DOS. **Productivity of popcorn and carioquinha beans intercropped under organic and mineral fertilization at different spacings.** 2008. 54f. Dissertation (Master's Degree in Soil and Water Management and Conservation) - Center for Agricultural Sciences, Federal University of Paraiba, Areia, 2008.

SEVERINO, F. J. **Suppression of weed infestation by the crop-livestock integration production system.** 2005. 113 f. Thesis (Doctorate in Plant Science) - Escola Superior de Agricultura "Luiz de Queiroz", Universidade de São Paulo, Piracicaba, 2005.

SHLINDWEIN, J. A. **Calibration of a method for determining and estimating doses of Phosphorus and Potassium in soil under a no-till system.** Porto Alegre, Federal University of Rio Grande do Sul, 2003. 169p. (Doctoral thesis).

SHUMAN, L. M. **Mineral Nutrition.** In: WILKINSON, R. E., ed. Plant-environment interactions. New York, Marcel Dekker, 1994. P. 149 - 182.

SILVA, A. J. DA.; UCHÔA, S. C. P.; ALVES, J. M. A.; LIMA, A. C. S.; SANTOS, C. S. V. DOS.; OLIVEIRA, J. M. F. DE.; MELO, V. F. Response of cowpea to doses and forms of phosphorus application in Yellow Latosol of the State of Roraima. **Revista Acta Amazônica,** v.40, n. 1, p. 31 - 36, 2010.

SILVA, C. C. da; SILVEIRA, P. M. da. Influence of agricultural systems on the response of irrigated bean (*Phaseolus vulgaris L.*) to nitrogen fertilization. **Pesquisa Agropecuària Tropical,** v.30,p.86-96, 2000.

SILVA, E. F. L.; ARAUJO, A. S. F.; SANTOS, V. B.; NUNES, L. A. P. L.; CARNEIRO. R. F. V. Biological fixation of N2 in cowpea under different doses and sources of soluble phosphorus. **Biosci. J.**, Umberlândia, v. 26, n. 3, p. 394 - 402, May/June 2010.

SILVA, J. A. **Initial application of P O₂₅ in the soil, evaluation in three successive crops in cowpea**. Areia: (Master's dissertation), UFPB-CCA, 2007. 53p.

SILVA, L. S. da; BOHNEN, H. Productivity and nutrient absorption by rice grown in nutrient solution with different levels of silicon and calcium. **Revista Brasileira Agrociência**, v. 9, n. 1, p. 49-52, 2003.

SILVA, M. G. *et al.* Soil management and nitrogen fertilization in winter bean cover. *In:*CONGRESSO NACIONAL DE PESQUISA DEFEIJÀO, 7, 2002a, Viçosa - MG. *Expanded Abstracts.***Viçosa**: UFV, 2002a. p. 612-614.

SILVA, M. G.; ARF, O.; SA, M. E.; RODRIGUES, R. A. F.; BUZETTI, S. nitrogen fertilization and soil management of winter common bean crop. **Scientia Agricola**, Piracicaba, v.61, n.3, p.307-312, 2004.

SILVA, P. S. L.; OLIVEIRA, C. N. Yields of green and mature beans from caupi cultivars. **Horticultura Brasileira**, Brasilia, v. 11, n. 2, p 133-135, 1993.

SILVA, T. R. B.; SORATTO, R. P.; CHIDI, S. N.; ARF, O.; SA, M. E. & BUZETTI, S. Doses and times of nitrogen application in winter bean cover crops. **Cult. Agron.**, 9:1-17, 2000.

SILVEIRA, P.M.; DAMASCENO, M. A. Studies of doses and installments of K and doses of N in the cultivation of irrigated beans. *In:* REUNIÂO NACIONAL DE PESQUISA DE FEIJÀO, 4., 1993, **Londrina**. *Abstracts...* Londrina: IAPAR, 1993. p.161.

SILVEIRA, P.M.; DAMASCENO, M.A. Doses and installments of K and N in irrigated bean crops. **Pesquisa Agropecuâria Brasileira**, v. 28, p. 1269-1276, 1993.

SMARTT, J. Grain legumes: evolution and genetic resources. Cambridge, Great Britain: **Cambridge University Press**, 1990, 333 p.

SMIDERLE, O. J. Nitrogen doses and physiological quality of cowpea seeds. **Roraima**: Embrapa Roraima, 2004.

SORATTO, R. P. Bean (*Phaseolus vulgaris* L.) response to nitrogen top dressing and molybdenum foliar application. II - Physiological quality of the seeds. In: REUNfAO NACIONAL DE PESQUISA DE FEIJÂO, 6, Salvador, 21/26 nov.1999.Anais. **Salvador**: EMBRAPA/CNPFA, v.1, p.595-598, 1999.

SORATTO, R. P.; CARVALHO, M. A. C. DE.; ARF, O. Nitrogen in cover crops for no-till bean cultivation. **Revista Brasileira de Ciência do Solo**, v. 30, p. 259-265, 2006.

SORATTO, R. P.; CARVALHO, M. A. C. & ARF, O. Chlorophyll content and bean productivity due to nitrogen fertilization. **Pesquisa Agropecuâria Brasileira**, v. 39, p.895-901, 2004.

SORATTO, R. P.; SILVA, T. R. B.; ARF, O. & CARVALHO, M. A. C. Levels and timing of nitrogen topdressing on no-till irrigated beans. **Cult. Agron.**, 10:89-99, 2002.

SOUZA, E. F. C.; SORATTO, R. P. Efeitos de fontes e doses de nitrogênio em cobertura, no-till corn. **Revista Brasileira de Milho e Sorgo**, Sete Lagoas, v.5 p.395-405, 2006.

SOUZA, R. F. de. **Phosphorus dynamics in soils under the influence of liming and organic fertilization, cultivated with beans**. 2005. 141 p. Thesis (Doctorate in Soils and Plant Nutrition) Federal University of Lavras, Lavras, 2005.

STONE, L. F.; MOREIRA, J. A. A. Bean plant response to nitrogen in top dressing, under different irrigation rates and soil preparations. **Pesquisa Agropecuâria Brasileira, Brasilia**, v.36, n.3, p.473-481, 2001.

TEDESCO, M. J., GIANELLO, C., BISSANI, C. A., BOHNEN, H. & VOLKWEISS, S. J. Análisis de solo, plantas e outros materiais. 2. ed. **Porto Alegre**, Universidade Federal do Rio Grande do Sul, 1995. 174p. (Technical bulletin, 5).

TEIXEIRA, I. R.; BORÉM, A.; SILVA, A. G.; KIKUTE, H. Zinc sources and doses in bean cultivated at different sowing times. **Acta Sci. Agron.** Maringâ, v. 30, n. 2, p. 255 - 259, 2008.

TISDALE, S. L,; NELSON, W. L,; BEATON, J. D. Soil fertility and fertilizers. 4. ed. **New York**: Macmillan, 1995. 754p.

TORRES S.B; OLIVEIRA F.N; OLIVEIRA R..C; FERNANDES JB. Productivity and morphology of caupi accessions in Mossoró, RN. **Horticultura Brasileira**, v. 26, n. 4, Oct.-Dec. 2008.

FEDERAL UNIVERSITY OF CEARA. Fertilization and liming recommendations for the state of Ceará. **Fortaleza**, UFC, 1993. 247p.

VALDERRAMA, M.; BUZETTI, S.; BENETT, C. G. S.; ANDREOTTI, M.; ARF, O.; SA, M. E. DE. Sources and doses of nitrogen and phosphorus in no-till bean crops. **Pesquisa Agropecuâria Tropical**, Goiânia, v. 39, n. 3, p. 191-196, jul./set. 2009.

VIEIRA, C. Cultivation of beans. **Viçosa-MG**: UFV, 1983. 146 p.

WILCOX, G. E.; FAGERIA, N. K. Nutritional deficiencies of beans, their identification and correction. **Goiânia**: Embrapa/CNPAF, 1976. 22 p. (Embrapa/CNPAF. Boletim, 5).

WOLFFENBUTTEL, R.; TEDESCO, M. J. Availability of sulphur for alfalfa in eight soils of Rio Grande do Sul and its relationship with soil parameters. **Agronomia Sul Rio Grandense**, Porto Alegre, v. 17, n. 2, p. 357 - 364, Jun. 1981.

XAVIER, T. F. efeito da adução nitrogenada sobre a nodulaçao do feijà caupi. **Paraiba**: Federal University of Piaui, 2006.

YOKOYAMA, L. P. **Cultivation of the common bean. Economic importance**, 2003. Available at: <http:/www.cnpaf.embrapa.br>, accessed on 23/01/2004.

Printed by Books on Demand GmbH, Norderstedt / Germany